A LOOK AT THE HUMAN BODY

THE MUSCULAR SYSTEM

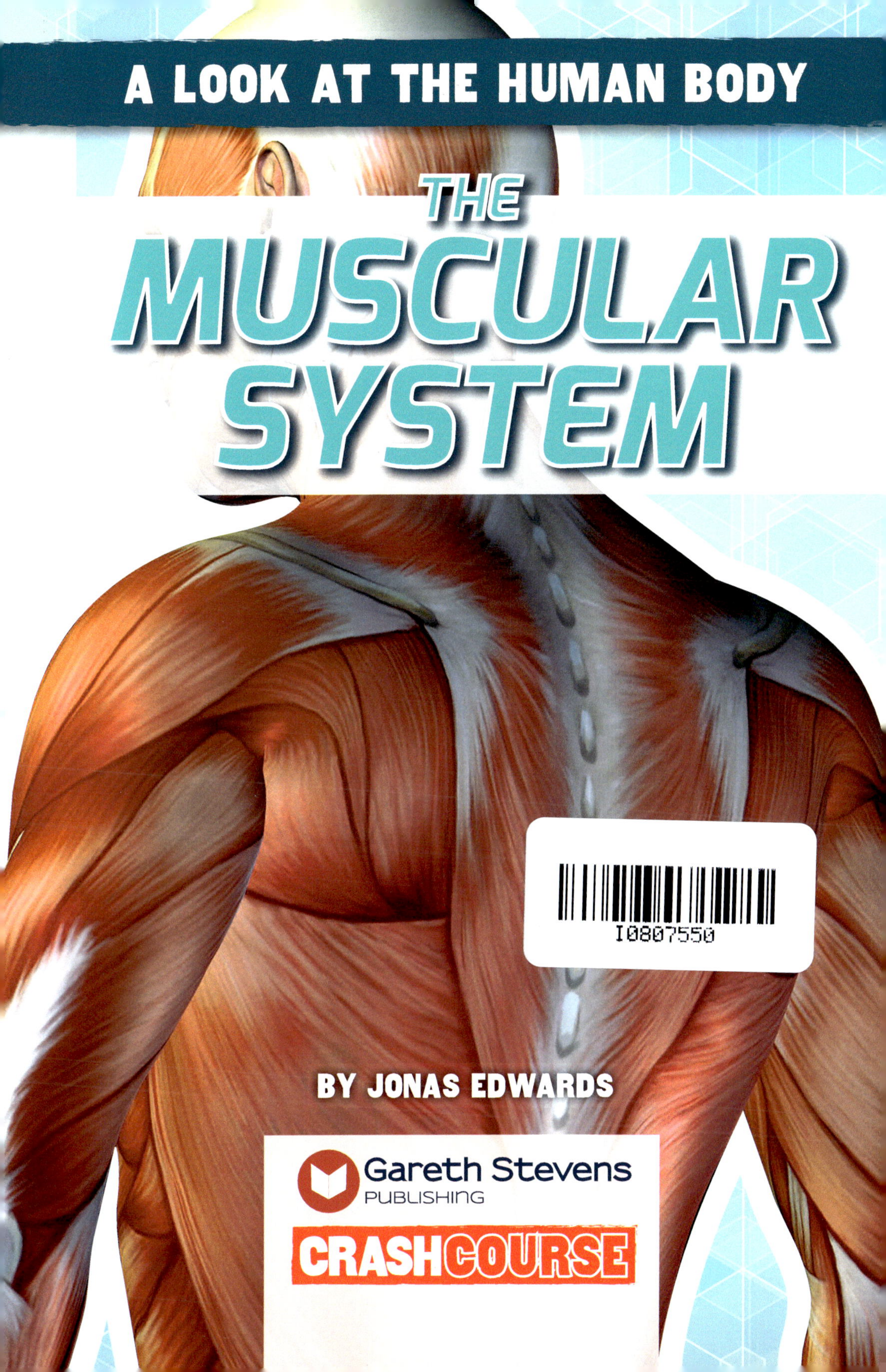

BY JONAS EDWARDS

Gareth Stevens PUBLISHING

CRASHCOURSE

Please visit our website, www.garethstevens.com. For a free color catalog of all our high-quality books, call toll free 1-800-542-2595 or fax 1-877-542-2596.

Library of Congress Cataloging-in-Publication Data

Names: Edwards, Jonas, author.
Title: The muscular system / Jonas Edwards.
Description: New York : Gareth Stevens Publishing, [2022] | Series: A look at the human body | Includes index. | Audience: Grades 2–3
Identifiers: LCCN 2020030131 | ISBN 9781538266199 (library binding) | ISBN 9781538266175 (paperback) | ISBN 9781538266182 (6 pack) | ISBN 9781538266205 (ebook)
Subjects: LCSH: Musculoskeletal system—Juvenile literature. | Muscles—Juvenile literature. | Human physiology—Juvenile literature.
Classification: LCC QP301 .E327 2022 | DDC 612.7/4—dc23
LC record available at https://lccn.loc.gov/2020030131

First Edition

Published in 2022 by
Gareth Stevens Publishing
29 E. 21st Street
New York, NY 10010

Editor: Kate Mikoley
Designer: Andrea Davison-Bartolotta

Photo credits: Cover design36/Shutterstock.com; background texture Yuliia Petrenko/Shutterstock.com; p. 5 Valeriy_G/iStock/Getty Images Plus/Getty Images; p. 7 (top) SCIEPRO/Science Photo Library/Getty Images; p. 7 (bottom) Teen00000/iStock/Getty Images Plus/Getty Images; p. 9 Dmytro Zinkevych/Shutterstock.com; p. 11 Sam Edwards/Stone/Getty Images; p. 13 Daisy-Daisy/iStock/Getty Images Plus/Getty Images; p. 15 FatCamera/E+/Getty Images; p. 17 (top) franckreporter/E+/Getty Images; p. 17 (bottom) BlueRingMedia/Shutterstock.com; p. 19 Alistair Berg/DigitalVision/Getty Images; p. 21 Janie Airey/Cultura/Getty Images; p. 23 Sebastian Kaulitzki/Science Photo Library/Getty Images; p. 25 (top) boonchai wedmakawand/Moment/Getty Images; p. 25 (bottom) Steve Gschmeissner/Science Photo Library/Getty Images; p. 27 (top) RgStudio/E+/Getty Images; p. 27 (bottom) Shubhangi Ganeshrao Kene/Science Photo Library; p. 29 Andersen Ross/The Image Bank/Getty Images; p. 30 Designua/Shutterstock.com.

Printed in the United States of America

CPSIA compliance information: Batch #CSGS22: For further information contact Gareth Stevens, New York, New York, at 1-800-542-2595.

CONTENTS

Words in the glossary appear in **bold** type the first time they are used in the text.

SO MANY MUSCLES!

Picture your muscles, the body parts that allow movement. Did you think about the ones you'd use to lift a heavy object, like the muscles in your arms? Those aren't the only muscles you have. Your body has more than 600 muscles!

MAKE THE GRADE

Lifting something heavy uses many muscles. The main muscles on the front of your upper arms are called biceps. If you lift something heavy, you might feel a bit of pain there.

MAKE A MUSCLE

Your muscles are made up of thousands of pieces of long, stringlike matter called fibers. Muscle **tissue** is somewhat stretchy, sort of like a rubber band. Muscles cause movement by getting smaller and tighter. This is called contracting.

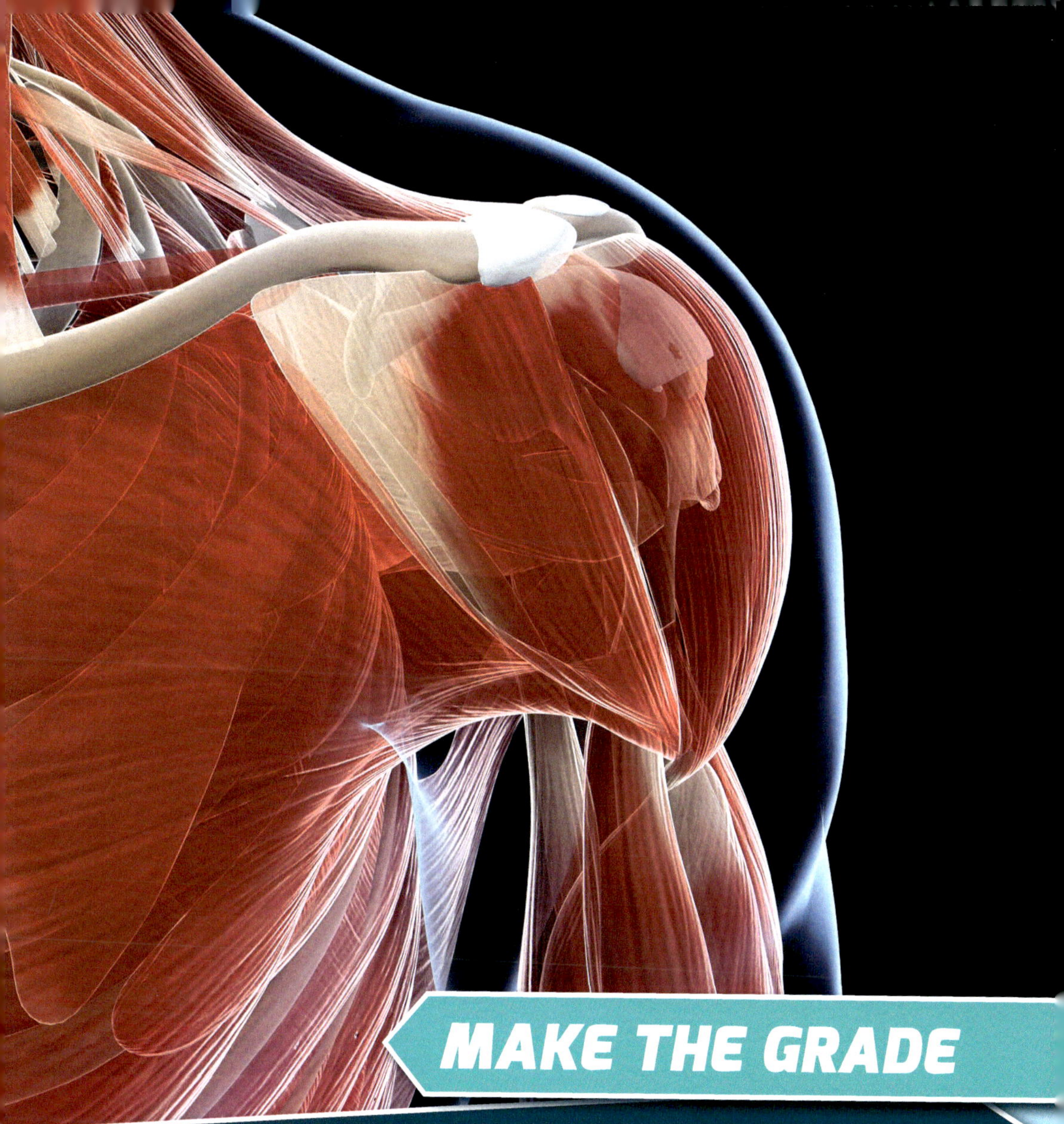

MAKE THE GRADE

Your brain tells your muscles when it's time to contract.

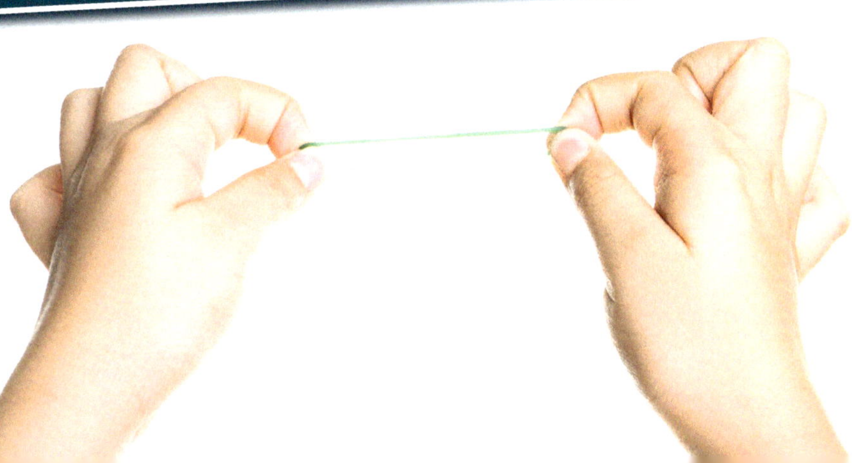

ANY VOLUNTEERS?

Some activities, such as walking or playing basketball, require you to think about using your muscles. For example, when you take a step, your legs move because you're choosing to move them. Muscles that work this way are called voluntary muscles.

MAKE THE GRADE

"Volunteer" means to choose to do something without being forced. That's where the name for voluntary muscles comes from—you can choose when and how to use them.

It might not always seem like you're thinking about moving your voluntary muscles. You're so used to walking, you probably don't think about using your **quadriceps**—you just do it! But your body doesn't just start walking on its own.

MAKE THE GRADE

You do some tasks with your voluntary muscles so often that your brain gets used to them. It might not seem like you're really controlling these movements, but you are!

DON'T THINK ABOUT IT!

Some muscles work without you even thinking about it. These are called involuntary muscles. They're muscles inside your **organs** that are controlled by a part of your brain you're not even aware of using!

MAKE THE GRADE

If you eat something bad and need to throw up, involuntary muscles push the food out of your stomach and out of your mouth.

With voluntary muscles, you choose when to make them move and when to make them stop. You can't do this with involuntary muscles. These muscles do their job whether you want them to or not. But you want them to—they help keep you alive!

MAKE THE GRADE

Your muscles account for about half of your body's weight!

THREE GROUPS

Your body has three different kinds of muscles. These are smooth, skeletal, and cardiac. Smooth and cardiac muscles are involuntary. Skeletal muscles are voluntary. Skeletal muscles are the muscles you likely think of when you think about being strong.

MAKE THE GRADE

Together with your bones, your skeletal muscles make up the musculoskeletal (musk-yoo-low-SKEL-uh-tul) system. This is what helps your body keep its form.

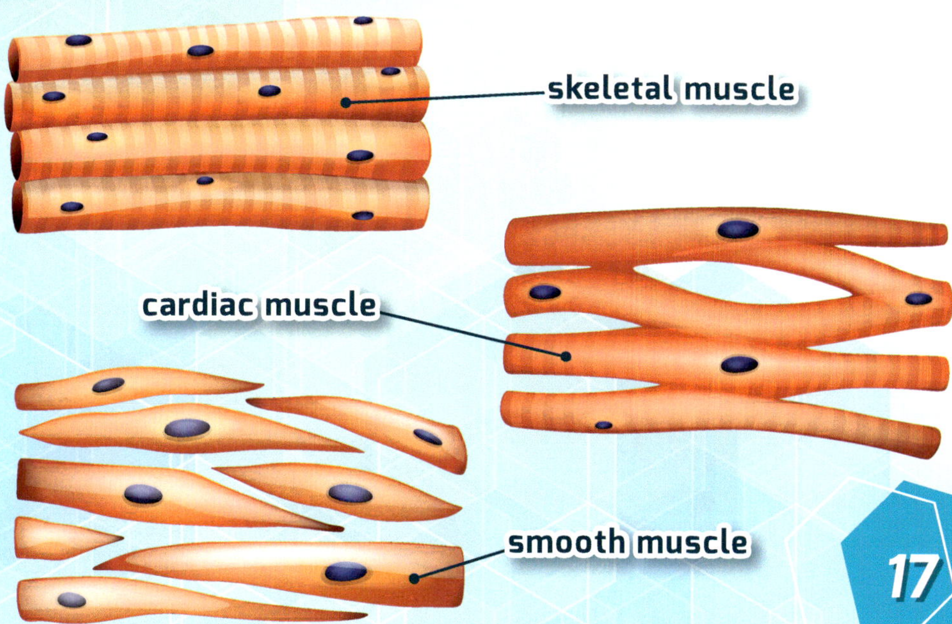

LET'S GET SKELETAL!

Your skeletal muscles are connected to your bones by special bands of tissue called tendons. Most skeletal muscles are attached on one end to one bone and on the other end to another bone. In the middle, the muscle stretches over a **joint**.

MAKE THE GRADE

When you contract a muscle, the connected bones move too. For example, when your hamstrings contract, bones in your leg move to make your knee bend.

You know that your arms and legs have skeletal muscles, but the rest of your body is covered with skeletal muscles too. For example, your abdominals, or abs, are muscles on the front of your belly. They help you sit up.

MAKE THE GRADE

There are muscles in your face too! You use these to make facial expressions, like raising your eyebrows, smiling, or frowning.

CARDIAC IN COMMAND

Cardiac muscle is found only in the heart. It keeps your heart beating and sends blood through your body. As this muscle contracts, it pumps blood out to the rest of your body. When it **relaxes**, blood can come back into your heart.

MAKE THE GRADE

Another name for cardiac muscle is myocardium (my-oh-KAR-dee-um).

SMOOTH MOVE!

Inside organs other than the heart, you can find smooth muscles. They're also found in **blood vessels**. Like cardiac muscle, these muscles are involuntary. While cardiac and skeletal muscles appear striped when looked at under a microscope, these muscles look smooth.

MAKE THE GRADE

Smooth muscles tend to come in layers, or sheets that lay one on top of the other.

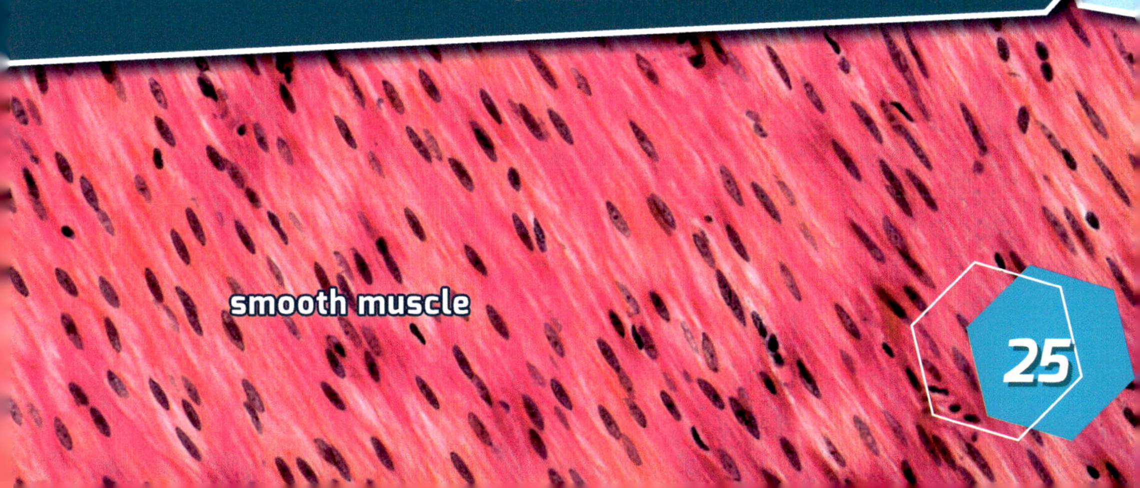

Smooth muscles help with many **functions** in your body, such as digestion, or the breakdown of the food you eat. Smooth muscles in your stomach and **intestines** help break down food so your body can use the **nutrients** in them.

MAKE THE GRADE

Smooth muscles work by contracting to move something through the organ. When muscles in your intestines contract, food is pushed through your body.

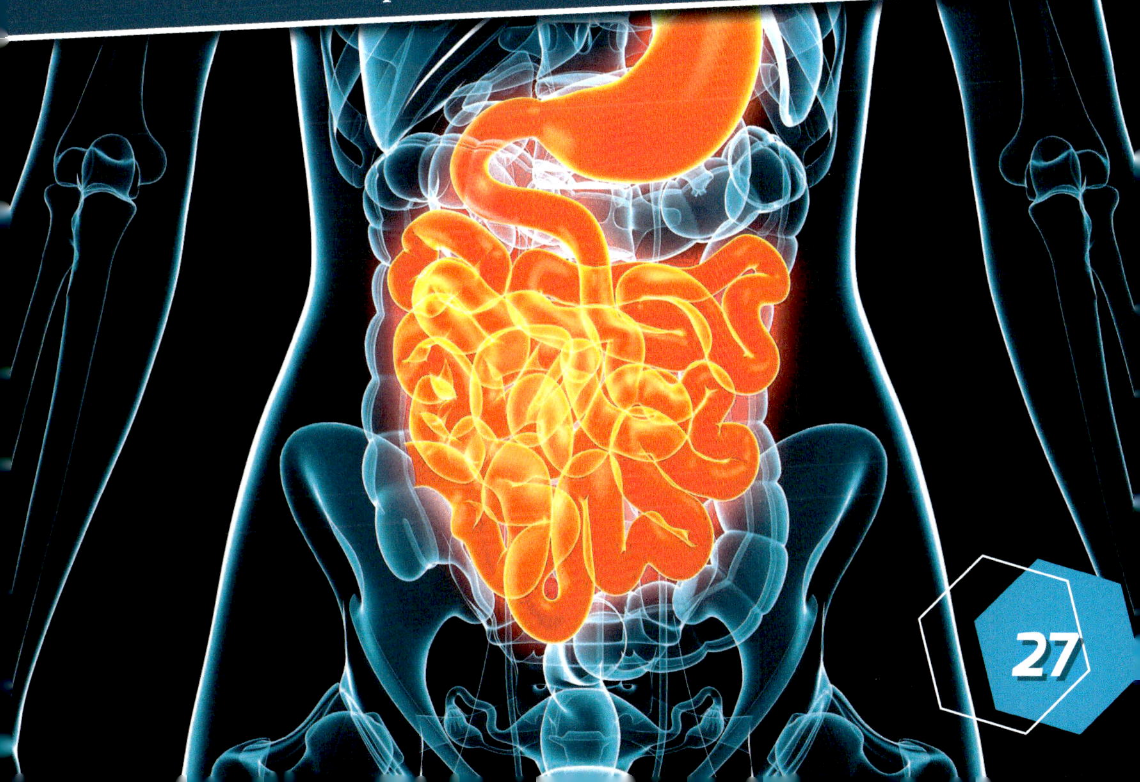

MUSCLES ON THE MOVE

Your muscular system is always working. Whether you're running, sitting, or sleeping, muscles in your body are hard at work. Eating healthy foods, exercising, and getting enough sleep are ways to make sure your muscular system stays healthy.

MAKE THE GRADE

Some exercises can make skeletal muscles bigger and stronger. This doesn't work for smooth muscles. Living a healthy life is the best way to keep all your muscles strong!

MAPPING THE MUSCLES

You have hundreds of muscles. Here are some of the ones discussed in this book.

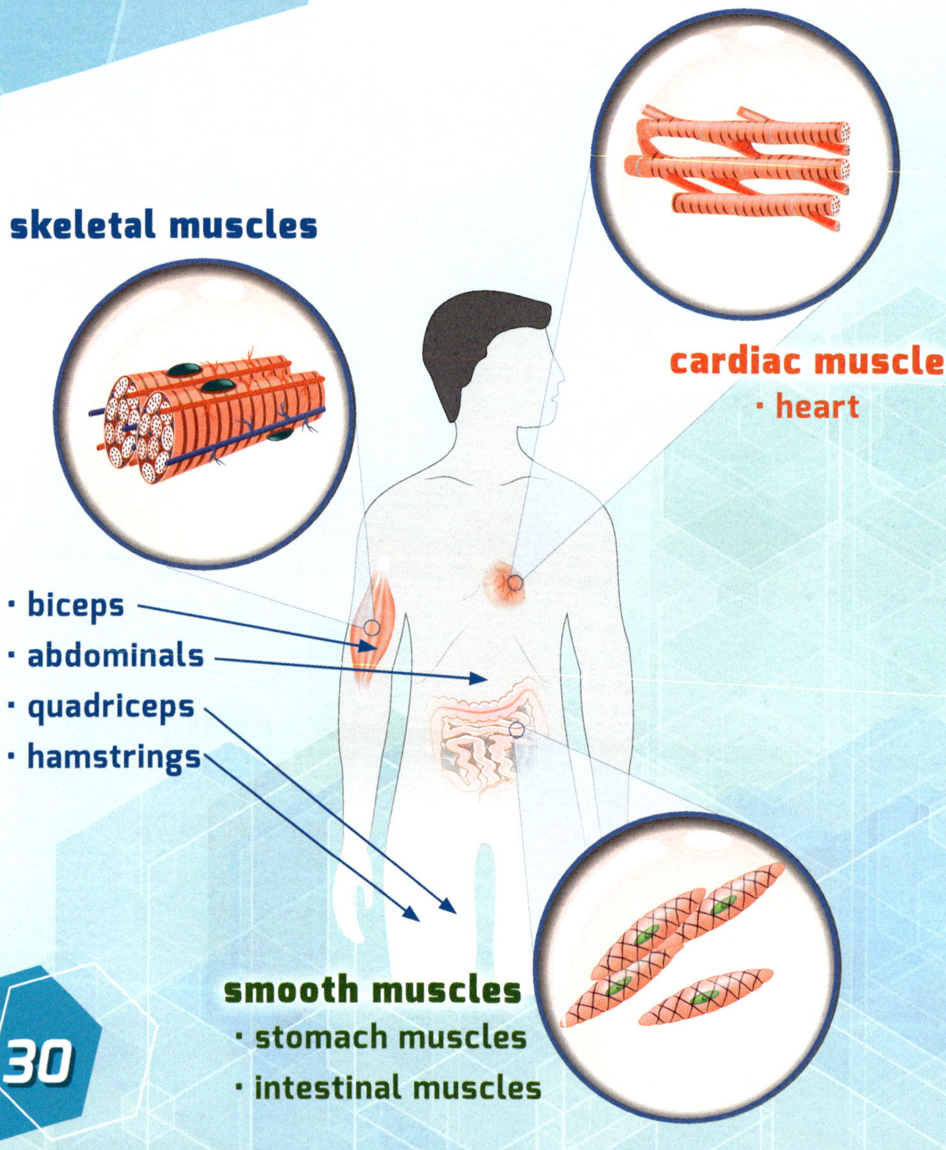

GLOSSARY

blood vessel: a small tube in a person's body that carries blood

facial expression: a look on a person's face

function: a special purpose or activity something carries out

hamstring: a muscle on the back of a person's upper leg

intestine: a tube in the body that helps digest food

joint: a point where two bones meet

nutrient: something a living thing needs to grow and stay alive

organ: a part inside a person's body

quadriceps: a large muscle at the front of the upper leg

relax: to become less tense or tight

tissue: matter that forms the parts of living things

FOR MORE INFORMATION

BOOKS

Beevor, Lucy. *Understanding Our Muscles*. Chicago, IL: Capstone Press, 2017.

Raij, Emily. *My Muscular System: A 4D Book*. North Mankato, MN: Pebble, 2019.

WEBSITES

Skeletal Muscles
www.dkfindout.com/us/human-body/muscles/skeletal-muscles/
Check out this interactive website to find out more about the body's skeletal muscles.

Your Muscles
kidshealth.org/en/kids/muscles.html
Go to this page to learn more about the hundreds of muscles in your body.

INDEX